CYBERSECURITY
Primer

Victor M. Font Jr.

A FontLife Publication

Cybersecurity Primer

ISBN: 978-1-62422-035-7 (print)
ISBN: 978-1-62422-033-3 (eBook)
Author: Victor M. Font Jr., Victor Font Consulting Group, LLC[1]

A FontLife Publication, LLC

Trademarks

Trademarked names may appear in this book. All terms mentioned in this book that are known to be trademarks or service marks have been appropriately capitalized. The author cannot attest to the accuracy of this information. Use of a term in this book should not be regarded as affecting the validity of any trademarks or service marks. Rather than use a trademark symbol with every occurrence of a trademarked name, we use the names only in an editorial fashion and to the benefit of the trademark owner, without intention of infringement of the trademark or service mark.

---

[1] https://victorfont.com/
[2] https://creativecommons.org/licenses/by/4.0/legalcode

Warning and Disclaimer

Co-Branding

Co-branding provides you with a cost-effective way of providing your customers with a uniquely styled gift. This title is available for co-branding, delivering a completely customized outer cover that can be tailored to match your corporate identity. For pricing, call us toll free at 1-844-842-3668 or write to info@victorfont.com.

TABLE OF CONTENTS

Introduction

Organizations rely heavily on the use of Information Technology (IT) products and services to run their day-to-day activities. Ensuring the security of these products and services is of the utmost importance for the success of the organization.

Today, Information Technology products and services face insidious threats from advanced malware and vulnerabilities that, if left unchecked, are designed to penetrate government, corporate, and infrastructure systems to gain control over those systems, rob unsuspecting victims, steal identities, damage reputations, hold us hostage, or worse.

Globally, Cybercrime damages are set to exceed $6 trillion each year[3] by 2021.

Despite the growing threat of Cyberattacks, more than half[4] of all businesses that suffered an attack didn't anticipate *any changes* to their security measures.

Ensuring the security of your IT assets is of the utmost importance for the success of your organization.

So how exactly do you prepare for the dismal future these statistics suggest?

This book serves as a starting-point for those new to Information Security and Cybersecurity. The intent is to provide a high-level overview of information/cyber security principles by introducing related concepts and the security control families that organizations can leverage to effectively secure their systems and information.

---

[3] https://cybersecurityventures.com/hackerpocalypse-cybercrime-report-2016/
[4] https://blog.barkly.com/cyber-security-statistics-2017

What's the Difference Between Information Security and Cybersecurity?

The difference between Information Security and Cybersecurity is a debate that rages on with as many different answers provided as the experts you query.

The terms "Cybersecurity" and "Information Security" are generally thought of as synonyms, but they create a lot of confusion even among security professionals. Some believe that Cybersecurity is a subset of Information Security while others think the opposite.

Yet, some banking regulators like the Reserve Bank of India, Hong Kong Monetary Authority, Monetary Authority of Singapore, etc., all require banks to have separate Cybersecurity and Information Security policies. These regulatory agencies view Cybersecurity and Information Security as two distinctly different objectives.

For the purposes of this book, we'll embrace the meanings of Cybersecurity and Information Security as defined by the National Institute of Standards and Technology (NIST)[5]:

Cybersecurity: The ability to protect or defend the use of cyberspace from cyberattacks.

Information Security: Protecting information and information systems from unauthorized access, use, disclosure, disruption, modification, or destruction in order to provide:

1) *confidentiality*, which means preserving authorized restrictions on access and disclosure, including means for protecting personal privacy and proprietary information;

2) *integrity*, which means guarding against improper information modification or destruction, and includes ensuring information non-repudiation and authenticity; and

---

[5] Source- https://nvlpubs.nist.gov/nistpubs/ir/2013/NIST.IR.7298r2.pdf

3) ***availability***, which means ensuring timely and reliable access to and use of information.

Information ≠ Data

To throw another log onto the fire, let's consider data security. Data security is all about securing data, but not every bit of data is information. So, what's the difference between data and information?

Data can be called information when it is interpreted within a context that gives it meaning. For example, "123-45-6789" is data because it's simply a string of alpha-numeric characters. If this data is found on a HR system record, then we know this is someone's social security number. Now it is information. Why? Because, it has context.

If we know this is someone's social security number, then it is information. In fact, it's personally identifiable information or PII, and that opens up a whole new can of worms. PII must be cybersecure. Significant fines and penalties can result when PII has been cyber-breached, especially in view of the new data privacy laws coming into effect.

To summarize:

- Information is data which has some meaning.
- Information Security is all about protecting the information, which focuses on its confidentiality, integrity, and availability (CIA).
- Cybersecurity is about protecting information from being launched into cyberspace through cyberattacks and breaches.

Target Audience

This book's target audience is anyone that wants to earn about Information Security principles and the precepts necessary to protect information and systems in a way that is commensurate with risk. This book provides a basic foundation of concepts and ideas to any person tasked with or interested in understanding how to secure systems.

For these reasons, this is a good resource for anyone seeking a better understanding of information and cybersecurity basics or a high-level view of the topic.

The Cybersecurity Ecosystem

Data breaches occur virtually on a daily basis. Cybercriminals are making enormous sums of money by selling breached data, information, or intellectual property on the dark web or in the gray or black markets. It's a multi-billion-dollar industry run by hacking operations often sponsored by nation-states, criminal organizations, or radical political groups.

Nation-state sponsored cyber-spies make it their full-time job to penetrate government, corporate, and infrastructure systems to gain control over those systems or collect intelligence to advance their own agendas. It's modern day espionage, but instead of inserting agents on the ground, intelligence and technology are compromised and accessed remotely from anywhere in the world.

Malware secretly installed on your computer can surreptitiously monitor your activity and steal your keystrokes, sending criminals your bank account and credit card login credentials, account numbers, social security number, or any other sensitive information you might otherwise transmit to an online destination.

Another type of malware can hijack your computer's CPU time and internet bandwidth to add your machine to a huge network of data mining systems that generate bitcoins and other cyber-currency variants. The only ones making money off of this is the criminals—*and you're paying for it!*

If you or your company has a website or online business, statistics show that more than half of the traffic hitting your server is coming from some cyber-bot or hack attempt. WordPress is a popular web application platform. Wordfence™, a premium security firewall for WordPress, is active on over 2-million WordPress powered sites and stops over 3.5M cyberattacks daily world-wide. Wordfence™ blocks over 1,500 attacks every month on this author's online business site.

The chances are good that if you've ever had an online account with any major retailer, credit card company, business, etc., there is some bit of information about you that can be found circulating on the dark web.

Top 10 Data Breaches of the 21st Century

Here is a list of the top 10 data breaches of the 21st century as compiled by CSOonline:

Organization	Impact
Details	
Yahoo	**3 billion user accounts**
Assumption: a state-sponsored actor hacked system. Compromised data includes names, dates of birth, email addresses, passwords, and security questions and answers.	
Adult Friend Finder	**More than 412.2 million accounts**
Hackers collected 20 years of data on six databases that included names, email addresses and passwords.	
eBay	**145 million users compromised**
Using the credentials of three corporate employees, hackers had full access for 229 days. Compromised data included names, addresses, dates of birth and encrypted passwords.	
Equifax	**143 million consumers**
Social Security Numbers, birth dates, addresses, and in some cases drivers' license numbers; 209,000 consumers also had their credit card data exposed.	
Heartland Payment Systems	**134 million credit cards**
Exposed through SQL injection used to install spyware on Heartland's data systems	
Target	**Credit/debit card information and/or contact information of up to 110 million people compromised**
Hackers gained access through a third-party HVAC vender to its point-of-sale (POS) payment card readers. The company estimated the cost of the breach at $162 million.	

Cybersecurity Primer

Organization	Impact
Details	
TJX Companies, Inc.	**94 million credit cards exposed**
Unknown how hackers compromised system. Two theories: breached wireless transfer between two Marshall's stores in Miami, Fla, or breaking into the TJX network through in-store kiosks that allow people to apply for jobs electronically	
Uber	**Personal information of 57 million Uber users and 600,000 drivers exposed**
Two hackers were able to access Uber's GitHub[6] account, where they found username and password credentials to Uber's Amazon Web Services account. Uber paid the hackers $100,000 to destroy the data with no way to verify that they did.	
JP Morgan Chase	**76 million households and 7 million small businesses**
Data included contact information—names, addresses, phone numbers and email addresses—as well as internal information about the users, according to a filing with the Securities and Exchange Commission[7]	
US Office of Personnel Management (OPM)	**Personal information of 22 million current and former federal employees and contractors**
Hackers, said to be from China, were inside the OPM system starting in 2012, but were not detected until March 20, 2014. A second hacker, or group, gained access to OPM through a third-party contractor in May 2014, but was not discovered until close to a year later. The intruders exfiltrated personal data—including in many cases detailed security clearance information and fingerprint data.	

---

[6] GitHub Inc. is a web-based hosting service for computer code version control. Microsoft is acquiring GitHub for US$7.5 billion.

[7] SEC filing: https://www.sec.gov/Archives/edgar/data/19617/000119312514362173/d799478d8k.htm

Organization	Impact
Details	
For details, see the House Committee on Oversight and Government Reform report: "The OPM Data Breach: How the Government Jeopardized Our National Security for More than a Generation."[8]	

According to the United Nations, the world population as of July 2018 stands at 7.6 billion. These 10 breaches represent over 4.2 billion compromised accounts. I know we're comparing apples to oranges, but if we could correlate a one-to-one relationship between breached accounts and an individual, 56% of all people have had their online identities compromised. That's a whole lot of cybercrime!

And these are just a handful of the breaches we know about so far. The breach list goes on and on and includes some major players like Sony's PlayStation Network, Anthem Blue Cross, RSA Security, Stuxnet, Verisign, Home Depot, Adobe, Experian, and many, many others. Some of these companies actually provide components for Cybersecurity implementations and they couldn't even protect themselves. Think about how many breaches there may be that haven't been discovered yet.

Stuxnet

Stuxnet is a different problem. Discovered in 2010, it is not a company, but is in fact malware in and of itself. It is a computer worm capable of transferring itself from device to device once it has penetrated a network. As far as can be determined, it was meant to attack Iran's nuclear power program, but it also serves as a template for real-world intrusion and service disruption of power grids, water supplies, or public transportation systems—aka the Internet of Things (IoT).

---

[8] https://oversight.house.gov/wp-content/uploads/2016/09/The-OPM-Data-Breach-How-the-Government-Jeopardized-Our-National-Security-for-More-than-a-Generation.pdf

Figure 1: Iranian President Mahmoud Ahmadinejad during a tour of uranium enrichment centrifuges at Natanz in 2008.
Source: Office of the Presidency of the Islamic Republic of Iran

Figure 2: Siemens Simatic S7-300 PLC CPU with three I/O modules attached
Image by Ulli1105 - Own work, CC BY-SA 2.5, https://commons.wikimedia.org/w/index.php?curid=16232 27

Stuxnet only targets Siemens Supervisory Control and Data Acquisition (SCADA) systems (above right). It damaged Iran's nuclear program by destroying an estimated 984 uranium enrichment centrifuges that were controlled by Siemens SCADA devices. Stuxnet spread quickly to Indonesia, India, Azerbaijan, United States, Pakistan, and other countries. The attack has been attributed to a joint effort by the US and Israel, although never officially acknowledged as such. If true, it means the worm couldn't be controlled and ironically turned on its own creators.

Cyber-MAD

The proliferation of Stuxnet adds a new threat to the world stage. During the cold war, the Governments of East vs. West kept each other at bay with a nuclear race that promised a Mutually Assured Destruction (MAD) for any nation launching the first strike. As a result of cold war propaganda, the worst fear for many people is the idea of a nuclear or electromagnetic pulse (EMP) attack that takes out a country's infrastructure or worse. Today, a greater threat exists then even nuclear annihilation. It's the modern equivalent of the cold war's MAD. It is Cyber-MAD, and there's no way of stopping it.

If or when there is a WWIII, it may very well begin with a Stuxnet attack on a country's power grid and infrastructure. When Stuxnet was released into the wild, it accidentally targeted and crippled critical infrastructure and industrial controls systems worldwide. In the consumer electronics world, it is the killer app. Everything in the modern world—kitchen appliances, ATMs, e-mail, home automation systems, electric grids, water infrastructure, oil and gas controls, communications and transportation systems—are all managed by industrial control systems.

For his book *Lights Out: A Cyberattack, A Nation Unprepared, Surviving the Aftermath*[9], author Ted Koppel interviewed the nation's top security experts. Koppel concluded:

> *"Top intelligence experts believe the Russians, Chinese, and Iranians are already 'inside the (U.S.) grid' with their versions of Stuxnet, and the only thing keeping them from destroying our systems is that the U.S. has cyberweapons in those nations' industrial control systems, too."*

Koppel further asserts that the federal government, while well prepared for natural disasters, has no plan for the aftermath of an attack on the power grid. A well-designed attack on just one of the nation's three electric power grids could cripple much of our infrastructure. It would create a blackout that lasts for weeks or months. Tens of millions of people across the United States are affected.

Without access to a generator and the fuel to run it, there is no running water, no sewage, no refrigeration or light. No fuel because pumps don't work. Food and medical supplies dwindle. The consumer electronic devices we rely on are dark. Banks no longer function, looting is widespread. The rule of law and order are challenged and stretched as never before.

Martial law is imposed. The military directly controls the normal civilian functions of government. And the best the current Secretary of Homeland Security can suggest is

---

[9] http://tedkoppellightsout.com/

keeping a battery-powered radio. One can only wonder what we'll be listening to if the radio stations are off the air because there's no power for their transmitters.

Intrusions by Any Other Name

Intrusions come in all sizes and shapes. We already discussed the Stuxnet computer worm, but intrusions go far beyond malware.

Intrusions are often targeted to compromise specific information. For example, customer information from Target and Anthem Blue Cross, intellectual property from Sony Pictures, and employee information from the OPM.

Another type of targeted attack is hacktivism. In February 2015, a hacktivist group calling themselves Lizard Squad hijacked Lenovo's website redirecting customers to a site that displayed selfie slideshows. It was an embarrassment for Lenovo and the resulting loss of revenue and remediation costs hurt their bottom line.

Intrusion Types/Methods

Phishing: A fraudulent attempt to obtain sensitive information such as usernames, passwords, credit card details, and money, often for malicious reasons, by disguising as a trustworthy entity in an electronic communication.

Phishing is typically carried out by email spoofing or instant messaging, directing users to enter personal information at a fake website, the look and feel of which are identical to the legitimate site. The only difference is the URL of the website in concern. An example might be an email from your credit card company announcing that your account has been locked. You are directed to login and correct the problem. Spoofing emails are often sent to distribution lists where the recipient names have been hidden. Hovering your mouse over the link without clicking it will reveal the link's destination.

Spear Phishing: A targeted phishing attempt directed at specific individuals or companies. Attackers may gather personal information about their target to increase their probability of success. This technique is by far the most successful on the Internet today, accounting for 91% of attacks[10].

Spear phishing is not always conducted through electronic communications. All that's required is for someone to click on a link. The link may even be on a social networking site like Facebook or Twitter.

Clone Phishing: A type of phishing attack where a legitimate and previously delivered email containing an attachment or link has had its content and recipient addresses hijacked and used to create an almost identical or cloned email. The attachment or link is replaced with a malicious version and sent from a spoofed email address to appear as though it came from the original sender.

Whaling: Phishing attacks directed specifically at senior executives and other high-profile targets.[11] Whaling scam emails masquerade as critical business communications, sent from a legitimate business authority. The content is often written as a legal subpoena, customer complaint, or executive issue.

DoS, DDoS: A denial-of-service attack (DoS) is a cyberattack in which the criminal seeks to make a machine or network resource unavailable to its intended users by temporarily or indefinitely disrupting the services of a host connected to the Internet. In a distributed denial-of-service attack (DDoS), the incoming traffic flooding the victim website originates from many different sources, effectively making it impossible to stop the attack by blocking a single source.

Botnet, Bots: A portmanteau of "robot" and "network", a botnet is a number of Internet-connected devices, each of which is running one or more bots. Botnets are used to perform DDoS attacks, steal data, send spam, and allow the attacker to access the target device and its connection.

---

[10] Stephenson, Debbie. "Spear Phishing: Who's Getting Caught?". Firmex. Retrieved July 31, 2018. (https://www.firmex.com/thedealroom/spear-phishing-whos-getting-caught/)
[11] "Fake subpoenas harpoon 2,100 corporate fat cats". The Register. Retrieved July 31, 2018. (https://www.theregister.co.uk/2008/04/16/whaling_expedition_continues/)

Threats and Vulnerabilities

A vulnerability is a weakness in a system, system security procedure, internal controls, or implementation that could be exploited by a threat source. Vulnerabilities leave systems susceptible to a multitude of activities that can result in significant and sometimes irreversible losses to an individual, group, or organization. These losses can range from a single damaged file on a laptop computer or mobile device to entire databases being compromised at an operations center. With the right tools and knowledge, an adversary can exploit system vulnerabilities and gain access to the information stored on them. The damage inflicted on compromised systems can vary depending on the threat source.

A threat source can be adversarial or non-adversarial. Adversarial threat sources are individuals, groups, organizations, or entities that seek to exploit an organization's dependence on cyber resources. Even employees, privileged users, and trusted users have been known to defraud organizational systems. Non-adversarial threat sources refer to natural disasters or erroneous actions taken by individuals in the course of executing their everyday responsibilities.

Fraud and Theft

Systems can be exploited for fraud and theft by "automating" traditional methods of fraud or by introducing new methods. System fraud and theft can be committed by insiders (i.e. authorized users) and outsiders. Authorized system administrators and users with access to and familiarity with the system (e.g., resources it controls, flaws) are often responsible for fraud. Former employees also pose a threat given their knowledge of the organization's operations, particularly if access is not terminated promptly.

Financial gain is one of the chief motivators behind fraud and theft, but financial systems are not the only systems at risk. There are several techniques that cybercriminals use to gather information they would otherwise not have had access to. Some of these techniques include:

Social Media: Social media (e.g., Facebook, Twitter, LinkedIn) has allowed cybercriminals to exploit the platform to conduct targeted attacks. Using easily-made, fake, and unverified social media accounts, cybercriminals can impersonate co-workers, customer service representatives, or other trusted individuals in order to send links to malicious code that steal personal or sensitive organizational information.

Social Engineering: In the context of Information Security, social engineering is a technique that relies heavily on human interaction to influence an individual to violate security protocol and encourages the individual to divulge confidential information. These types of attacks are commonly committed via phone or online. Attacks perpetrated over the phone are the most basic social engineering attacks being committed. For example, an attacker may mislead a company into believing the attacker is an existing customer and have that company divulge information about that customer.

Advanced Persistent Threat (APT): A set of stealthy and continuous computer hacking processes, often orchestrated by a person or persons targeting a specific entity. APTs generally target private organizations, states, or both for business or political motives. APT processes require a high degree of covertness over a long period of time. "Advanced" signifies sophisticated techniques using malware to exploit vulnerabilities in systems. "Persistent" suggests an external command and control system is continuously monitoring and extracting data from a specific target. "Threat" indicates human involvement in orchestrating the attack.

Insider Threat

Employees can represent an insider threat to an organization given their familiarity with the employer's systems and applications as well as what actions may cause the most damage, mischief, or disorder. Employee sabotage—often instigated by knowledge or threat of termination—is a critical issue for organizations and their systems. In an effort to mitigate the potential damage caused by employee sabotage, the terminated employee's access to IT infrastructure should be immediately disabled, and the individual should be escorted off company premises.

Examples of system-related employee sabotage include, but are not limited to:

- Destroying hardware or facilities;
- Planting malicious code that destroys programs or data;
- Entering data incorrectly, holding data, or deleting data;
- Crashing systems; and
- Changing administrative passwords to prevent system access.

Malicious Hackers and Their Motivations

Malicious hacker is a term used to describe an individual or group who use an understanding of systems, networking, and programming to illegally access systems, cause damage, or steal information. Understanding the motivation that drives a malicious hacker can help an organization implement the proper security controls to prevent the likelihood of a system breach.

Attackers: Attackers break into networks for the thrill and challenge or for bragging rights in the attacker community. While remote hacking once required considerable skills or computer knowledge, attackers can now download attack scripts and protocols from the Internet and launch them against victim sites.

Bot-Network Operators: Bot-network operators assume control of multiple systems to coordinate attacks and distribute phishing schemes, spam, and malicious code. The services of compromised systems and networks can be found in underground markets online (e.g., purchasing a denial of service attack, using servers to relay spam, or phishing attacks).

Criminal Groups: Criminal groups seek to attack systems for monetary gain. Specifically, organized crime groups use spam, phishing, and spyware/malicious code to commit identity theft and online fraud. International corporate spies and organized crime organizations also pose threats to the Nation based on their ability to conduct industrial espionage, large-scale monetary theft, and the recruitment of new attackers. Some criminal groups may try to extort money from an organization by threatening a cyberattack or by encrypting and disrupting its systems for ransom.

Foreign Intelligence Services: Foreign intelligence services use cyber tools as part of their information gathering and espionage activities. In addition, several nations are aggressively working to develop information warfare doctrines, programs, and capabilities. Such capabilities enable a single entity to have a significant and serious impact by disrupting the supply, communications, and economic infrastructures that support military power—impacts that could affect the daily lives of U.S. citizens.

Phishers: Phishers are individuals or small groups that execute phishing schemes to steal identities or information for monetary gain. Phishers may also use spam and spyware/malicious code to accomplish their objectives.

Spammers: Spammers are individuals or organizations that distribute unsolicited e-mail with hidden or false information to sell products, conduct phishing schemes, distribute spyware/malicious code, or attack organizations (e.g., DoS).

Spyware/Malicious Code Authors: Individuals or organizations who maliciously carry out attacks against users by producing and distributing spyware and malicious code. Destructive computer viruses and worms that have harmed files and hard drives include the Melissa Macro Virus, the Explore.Zip worm, the CIH (Chernobyl) Virus, Nimda, Code Red, Slammer, and Blaster.

Terrorists: Terrorists seek to destroy, incapacitate, or exploit critical infrastructures to threaten national security, cause mass casualties, weaken the U.S. economy, and damage public morale and confidence. Terrorists may use phishing schemes or spyware/malicious code to generate funds or gather sensitive information. They may also attack one target to divert attention or resources from other targets.

Industrial Spies: Industrial espionage seeks to acquire intellectual property and know-how using clandestine methods.

Non-Adversarial Threat Sources and Events

Errors and Omissions: Errors and omissions can be inadvertently caused by system operators who process hundreds of transactions daily or by users who create and edit data on organizational systems. Errors by users, system operators, or programmers may occur throughout the life cycle of a system and may directly or indirectly contribute to security problems, degrade data and system integrity. Software applications, regardless of the level of sophistication, are not capable of detecting all types of input errors and omissions. Therefore, it is the responsibility of the organization to establish a sound awareness and training program to reduce the number and severity of errors and omissions.

Loss of Physical and Infrastructure Support: The loss of supporting infrastructure includes power failures (e.g., outages, spikes, brownouts), loss of communications, water outages and leaks, sewer malfunctions, disruption of transportation services, fire, flood, civil unrest, and strikes. A loss of supporting infrastructure often results in system downtime in unexpected ways. For example, employees may not be able to get to work during a winter storm, although the systems at the work site may be functioning as normal.

Information Sharing and the Impacts to Personal Privacy: The accumulation of vast amounts of personally identifiable information by government and private organizations has created numerous opportunities for individuals to experience privacy problems as a byproduct or unintended consequence of a breach in security.

Individuals' voluntarily sharing PII through social media has also contributed to new threats that allow malicious hackers to use that information for social engineering or to bypass common authentication measures.

Organizations may share information about cyber threats that includes PII. These disclosures could lead to unanticipated uses of such information, including surveillance or other law enforcement actions.

The Role of Malicious Code in Cyberattacks

Malicious code refers to viruses, Trojan horses, worms, logic bombs, and any other software created for the purpose of attacking a platform.

Virus: A code segment that replicates by attaching copies of itself to existing executables. The new copy of the virus is executed when a user executes the new host program. The virus may include an additional "payload" that triggers when specific conditions are met.

Trojan Horse: A program that performs a desired task, but that also includes unexpected and undesirable functions. For example, consider an editing program for a multiuser system. This program could be modified to randomly and unexpectedly delete a user's files each time they perform a useful function (e.g., editing).

Worm: A self-replicating program that is self-contained and does not require a host program or user intervention. Worms commonly utilize network services to propagate to other host systems.

Logic Bomb: This type of malicious code is a set of instructions secretly and intentionally inserted into a program or software system to carry out a malicious function at a predisposed time and date or when a specific condition is met.

Ransomware: Is a type of malicious code that blocks or limits access to a system by locking the entire screen or by locking down or encrypting specific files until a ransom is paid. There are two different types of ransomware attacks—encryptors and lockers:

- Encryptors block (encrypt) system files and demand a payment to unblock (or decrypt) those files. Encryptors, or crypto-ransomware, are the most common and most worrisome (e.g., WannaCry).

- Lockers are designed to lock users out of operating systems. The user still has access to the device and other files, but in order to unlock the infected computer, the user is asked to pay a ransom. To make matters worse, even if the user pays the ransom, there is no guarantee that the attacker will actually provide the decryption key or unlock the infected system.

Anatomy of a Cyberattack

A cyberattack is any attempt to expose, alter, disable, destroy, steal or gain unauthorized access to or make unauthorized use of an asset by hacking into a susceptible system.

Cyberattacks are offensive maneuvers that target computer information systems, infrastructures, computer networks, or personal computer devices. They could be prosecuted by nation-states, individuals, groups, society, or organizations. They may originate from anonymous sources.

Cyberattacks are conducted as cybercampaigns, cyberwarfare, or cyberterrorism depending on the context. Cyberattacks can vary in scope and range from installing spyware on a single personal computer to attempting to destroy the infrastructure of an entire nation. Cyberattacks have become increasingly sophisticated and dangerous.

Cyberwarfare utilizes techniques of defending and attacking information and computer networks that inhabit cyberspace, often through a prolonged cybercampaign or series of related campaigns. It denies an opponent's ability to do the same, while employing technological instruments of war to attack an opponent's critical computer systems.

Cyberterrorism is "the use of computer network tools to shut down critical national infrastructures (such as energy, transportation, government operations) or to coerce or intimidate a government or civilian population".[12]

The end result of both cyberwarfare and cyberterrorism is the same—to damage critical infrastructures and computer systems linked together within the confines of cyberspace.

---

[12] Lewis, James. United States. Center for Strategic and International Studies. Assessing the Risks of Cyber Terrorism, Cyber War and Other Cyber Threats. Washington, D.C., 2002.

Battle Maneuvers

Just as soldiers conduct certain strategies during a conflict, cybercriminals employ battle tactics and stratagems as well. They will do anything necessary to gain tactical or strategic advantage by accessing a system once they decide it will be profitable, challenging, or fun for them to do so. Battle tactics include:

1. **Reconnaissance**: Cybercriminals reconnoiter their victims and plan their attacks. They research, identify, and select targets by phishing, harvesting email addresses, engaging in social engineering, and other sneaky tactics. They also use various tools to scan and exploit network vulnerabilities, services, and applications.

2. **Weaponization and Payload Delivery**: Next, the attackers choose their weapon (malware payload) and the delivery vehicle:

   a. A *drive-by download* delivers an exploit or advanced malware covertly, usually by taking advantage of a vulnerability in a web browser, operating system, or third-party application

3. **Exploitation**: An attacker generally has two options for exploitation:

   a. *Social engineering* (as previously discussed), and

   b. *Software exploits*—a more sophisticated technique that tricks the web browser, operating system, or other third-party software into executing an attacker's code.

   Exploits have become an efficient and stealthy method to deliver advanced malware to infiltrate a network or system because they can be hidden in legitimate files. Once the exploitation has succeeded, an advanced malware payload can be installed.

4. **Installation**: After a targeted endpoint is infiltrated, the attacker needs to ensure survivability. Various types of advanced malware are used for resilience or persistence, including:

a. *Rootkits* provide privileged root-level access to a computer

b. *Bootkits* are kernel mode variants of rootkits ordinarily used to attack computers that are protected by full-disk encryption

c. *Backdoors* are often installed as a failover to enable an attacker to bypass normal authentication procedures in order to gain access to a compromised system in case the primary payload is detected and removed from the system.

d. *Anti-AV* software disables any legitimately installed antivirus software on the compromised endpoint. This prevents the automatic detection and removal of malware. Many anti-AV programs infect the master boot record (MBR) of the target endpoint.

5. ***Command and Control (CnC)***: Communication is the lifeblood of a successful attack. Attackers must maintain communications with infected systems to effectuate command and control, and to retrieve data stolen from a target system or network.

CnC communications are clandestine. They can't raise any suspicion on the network. Such traffic is usually silenced through obfuscation or hidden through techniques that include:

a. *Encryption* with SSL, SSH, some other custom application, or proprietary encryption. BitTorrent is known for its proprietary encryption. It's a favorite tool both for injecting infections and CnC.

b. *Circumvention* via proxies, remote access tools, or by tunneling, which is a communications protocol that allows for the secure movement of data from one network to another through a process called encapsulation.

c. *Port evasion* using network anonymizers or port hopping to tunnel over open or nonstandard ports.

   d. *Fast flux (dynamic DNS)* to proxy through multiple infected hosts, reroute traffic, and make it extremely difficult for forensic teams to figure out where traffic is really going.

6. ***Playing the Long Game***: As we previously learned, attackers have many different motives for their actions. Attacks can often last months or even years, particularly when the objective is data theft, where the attacker plays the long game and uses a low, slow, fly under the radar attack strategy to avoid detection.

Cornerstones of Information Security

There are eight cornerstones in Information Security:

1. Information Security supports the mission of the organization.
2. Information Security is an integral element of sound management.[13]
3. Information Security protections are implemented so as to be commensurate with risk.
4. Information Security roles and responsibilities are explicit.
5. Information Security responsibilities for system owners go beyond their own organization.
6. Information Security requires a comprehensive and integrated approach.
7. Information Security is assessed and monitored regularly.
8. Information Security is constrained by societal and cultural factors.

---

[13] sound management refers to due diligence in taking all practical steps to ensure that Information Security management decisions are made in such a way that they not only protect the information stored, processed, and transmitted by an organization, but also the systems that fall under the purview of the organization.

Information Security Supports the Mission of the Organization

Information Security is defined as the protection of information and systems from unauthorized access, use, disclosure, disruption, modification, or destruction in order to provide confidentiality, integrity, and availability. The careful implementation of Information Security controls[14] is vital to protecting an organization's information assets as well as its reputation, legal position, personnel, and other tangible or intangible assets.

Information Security Is an Integral Element of Sound Management

Management personnel are ultimately responsible for determining the level of acceptable risk for a specific system and the organization as a whole, while considering the cost of security controls. Since Information Security risk cannot be completely eliminated, the objective is to find the optimal balance between protecting the information or system and utilizing available resources. It is vital for systems and related processes to have the ability to protect information, financial assets, physical assets, and employees, while also taking resource availability into consideration.

Information Security Protections Are Implemented So as To Be Commensurate with Risk

Risk to a system can never be completely eliminated. Therefore, it is crucial to manage risk by striking a balance between usability and the implementation of security protections. The primary objective of risk management is to implement security protections that are commensurate with risk. Applying unnecessary protections may waste resources and make systems more difficult to use and maintain. Conversely, not applying measures needed to protect the system may leave it and its information vulnerable to breaches in

---

[14] Amazon Web Services has an excellent example of a NIST based security controls matrix in their Standardized Architecture for NIST-based Assurance Frameworks on the AWS Cloud: Quick Start Reference Deployment (https://docs.aws.amazon.com/quickstart/latest/compliance-nist/welcome.html). Download the AWS security controls matrix here: https://fwd.aws/bWvRw

confidentiality, integrity, and availability, all of which could impede or even halt the mission of the organization.

Information Security Roles and Responsibilities Are Explicit

The roles and responsibilities of system owners, common control providers, authorizing officials, system security officers, users, and others are clear and documented. If the responsibilities are not made explicit, management may find it difficult to hold personnel accountable for future outcomes.

Information Security Responsibilities for System Owners Go Beyond Their Own Organization

Users of a system are not always located within the boundary of the system they use or have access to. For example, when an interconnection between two or more systems is in place, Information Security responsibilities might be shared among the participating organizations. When such is the case, the system owners are responsible for sharing the security measures used by the organization to provide confidence to the user that the system is adequately secure and capable of meeting security requirements. In addition to sharing security-related information, the incident response team has a duty to respond to security incidents in a timely fashion in order to prevent damage to the organization, personnel, and other organizations.

Information Security Requires A Comprehensive and Integrated Approach

Providing effective Information Security requires a comprehensive approach that considers a variety of areas both within and outside of the Information Security field. This approach applies throughout the entire system life cycle.

Security controls are seldom put in place as stand-alone solutions to a problem. They are typically more effective when paired with another control or set of controls. Security controls, when selected properly, can have a synergistic effect on the overall security of a system. Each security control has a related controls section listing security control(s) that compliment that specific control. If users do not understand these interdependencies, the results can be detrimental to the system.

Interdependencies between and amongst security controls are not the only factor that can influence the effectiveness of security controls. System management, legal constraints, quality assurance, privacy concerns, and internal and management controls can also affect the functionality of the selected controls. System managers must be able to recognize how Information Security relates to other security disciplines like physical and environmental security.

Information Security Is Assessed and Monitored Regularly

Information Security is not a static process and requires continuous monitoring and management to protect the confidentiality, integrity, and availability of information as well as to ensure that new vulnerabilities and evolving threats are quickly identified and responded to accordingly. In the presence of a constantly evolving workforce and technological environment it is essential that organizations provide timely and accurate information while operating at an acceptable level of risk.

Information Security Is Constrained by Societal and Cultural Factors

Societal factors influence how individuals understand and use systems which consequently impacts the Information Security of the system and organization. Individuals perceive, reason, and make risk-based decisions in different ways. To address this, organizations make Information Security functions transparent, easy to use, and understandable. Additionally, providing regularly scheduled security awareness training mitigates individual differences of risk perception.

As with societal factors, how an organization conducts business can serve as a cultural factor worth considering when dealing with Information Security. An organization's own culture can impact its response to Information Security. Careful explanation of the risks associated with the business practices can help in the transparency and acceptance of the recommended Information Security practices.

Information Security Policy

Information Security policy is defined as an aggregate of directives, regulations, rules, and practices that prescribes how an organization manages, protects, and distributes information. The term policy can also refer to specific security rules for a system or even the specific managerial decisions that dictate an organization's email privacy policy or remote access security policy.

Standards, Guidelines, and Procedures

Because policy is written at a broad level, organizations develop standards, guidelines, and procedures that offer users, managers, system administrators, and others a clear approach to implementing policy and meeting organizational goals.

Standards and guidelines specify technologies and methodologies to be used to secure systems. Procedures are more detailed steps to be followed to accomplish security-related tasks. Standards, guidelines, and procedures may be disseminated throughout an organization via handbooks, regulations, manuals, or computer based or classroom training.

- Organizational standards specify uniform use of specific technologies, parameters, or procedures when it will benefit an organization. Standardization of organization-wide identification badges is a typical example, providing ease of employee mobility and automation of entry/exit systems. Standards are normally compulsory within an organization.

- Guidelines assist users, systems personnel, and others in effectively securing their systems. The nature of guidelines, however, immediately recognizes that systems vary considerably, and imposition of standards is not always achievable, appropriate, or cost effective.

- Procedures describe how to implement applicable security policies, standards, and guidelines. They are detailed steps to be followed by users, system operations personnel, or others to accomplish a particular objective.

Some organizations issue overall Information Security manuals, regulations, handbooks, or similar documents. These may mix policy, guidelines, standards, and procedures, since they are closely linked. While manuals and regulations can serve as important tools, it is often useful if they clearly distinguish between policy and its implementation. This can help in promoting flexibility and cost-effectiveness by offering alternative implementation approaches to achieving policy goals.

Program Policy

Program policy is used to create an organization's Information Security program. Program policies set the strategic direction for security and assign resources for its implementation within the organization. A management official—typically the Chief Information Security Officer (CISO)—issues program policy to establish or restructure the organization's Information Security program. This high-level policy defines the purpose of the program and its scope within the organization, addresses compliance issues, and assigns responsibility to the Information Security organization for direct program implementation as well as other related responsibilities.

Issue-Specific Policy

Based on the guidance from the Information Security policy, issue-specific policies are developed to address areas of current relevance and concern to an organization. The intent is to provide specific guidance and instructions on proper usage of systems to employees within the organization. An issue-specific policy is meant for every technology the organization uses and is written in such a way that it will be clear to users. Unlike program policies, issue-specific policies must be reviewed on a regular basis due to frequent technological changes in an organization.

System-Specific Policy

Program and issue-specific policies are broad, high-level policies written to encompass the entire organization where system-specific policies provide information and direction on what actions are permitted on a particular system. These policies are similar to issue-specific policies in that they relate to specific technologies throughout the organization. However, system-specific policies dictate the appropriate security configurations to the personnel responsible for implementing the required security controls in order to meet the organization's Information Security needs.

Interdependencies

Policy is related to many of the topics covered in this publication:

- **Program Management**: Policy is used to establish an organization's Information Security program and is therefore closely tied to program management and administration. Both program and system-specific policy may be established in any of the areas covered in this publication. For example, an organization may wish to have a consistent approach to contingency planning for all its systems and would issue appropriate program policy to do so. On the other hand, it may decide that its systems are sufficiently independent of each other that system owners can deal with incidents on an individual basis.

- **Access Controls**: System-specific policy is often implemented using access controls. For example, it may be a policy decision that only two individuals in an organization are authorized to run a check-printing program. Access controls are used by the system to implement or enforce this policy.

- **Links to Broader Organizational Policies**: It is important to understand that Information Security policies are often extensions of other organizational policies. Support and coordination should be reciprocal between Information Security and other organizational policies to minimize confusion. For example, an organization's email policy would likely be relevant to its broader policy on privacy.

Cost Considerations

A number of potential costs are associated with developing and implementing Information Security policies. The most significant costs are implementing the policy and addressing its subsequent impacts on the organization, its resources, and personnel. The establishment of an Information Security program, accomplished through policy, does not come at a negligible cost.

Risk Management

Risk is a measure of the extent an entity is threatened by a potential circumstance or event, and typically a function of:

(i)   the adverse impacts that would arise if the circumstance or event occurs; and

(ii)   the likelihood of occurrence.

Individuals manage risks every day, though they may not be aware of it. Actions as routine as buckling a car safety belt, carrying an umbrella when rain is forecast, or writing down a to-do list rather than trusting memory all fall under the purview of risk management. Individuals recognize various threats to their best interests and take precautions to guard against them or to minimize their effects.

With respect to Information Security, risk management is the process of minimizing risks to organizational operations (e.g., mission, functions, image, and reputation), organizational assets, individuals, other organizations, and the Nation resulting from the operation of a system. There are four distinct steps for risk management. Risk management requires organizations to (i) frame risk, (ii) assess risk, (iii) respond to risk, and (iv) monitor risk.

(i)   *Framing Risk*: how organizations establish a risk context for the environment in which risk-based decisions are made. The purpose of the risk framing component is to produce a risk management strategy that addresses how organizations intend to assess, respond to, and monitor risk—while making explicit and transparent the risk perceptions that organizations routinely use in making both investment and operational decisions.

(ii)   *Assessing Risk*: how organizations analyze risk within the context of the organizational risk frame. The purpose of the risk assessment component is to identify: (i) threats to organizations operations and assets, individuals, other organizations, and the Nation; (ii) internal and external vulnerabilities of organizations; (iii) the harm (i.e., consequences/impact) to organizations that may occur given the potential for threats exploiting vulnerabilities; and (iv) the likelihood that harm will occur.

(iii)   ***Responding to Risk***: how organizations respond to risk once that risk is determined based on the results of risk assessments. The purpose of the risk response component is to provide a consistent, organization-wide response to risk in accordance with the organizational risk frame by: (i) developing alternative courses of action for responding to risk; (ii) evaluating the alternative courses of action; (iii) determining appropriate courses of action consistent with organizational risk tolerance; and (iv) implementing risk responses based on selected courses of action.

(iv)   ***Monitoring Risk***: how organizations monitor risk over time. The purpose of the risk monitoring component is to: (i) verify that planned risk response measures are implemented and that Information Security requirements derived from/traceable to organizational missions/business functions, legislation, directives, regulations, policies, standards, and guidelines are satisfied; (ii) determine the ongoing effectiveness of risk response measures following implementation; and (iii) identify risk-impacting changes to organizational systems and the environments in which the systems operate.

Assurance

Information assurance is the degree of confidence one has that security measures protect and defend information and systems by ensuring their availability, integrity, authentication, confidentiality, and non-repudiation. These measures include providing for restoration of systems by incorporating protection, detection, and reaction capabilities.

Assurance is not, however, an absolute guarantee that the measures will work as intended. Understanding this distinction is crucial as quantifying the security of a system can be daunting. Nevertheless, it is something individuals expect and obtain, often without realizing it. For example, an individual may routinely receive product recommendations from colleagues but may not consider such recommendations as providing assurance.

There are two categories of assurance methods and tools to consider in your planning: the design and subsequent implementation of assurance and operational assurance (further categorized into audits and monitoring). The division between the two categories can be ambiguous at times as there is significant overlap. While such issues as configuration management or audits are generally included in operational assurance, they may also be vital during a system's development. This tends to focus more on technical issues during design and implementation assurance and is a mixture of management, operational, and technical issues under operational assurance. At a high-level, Assurance includes the following:

1. *Authorization*: the official management decision to authorize the operation of a system.

   1.1. *Authorization and Assurance*: Assurance is an integral element in making the decision to authorize a system to operate. Assurance addresses whether the technical measures and procedures are operating according to a set of security requirements and specifications as well as general quality principles.

   1.2. *Authorization of Products to Operate in a Similar Situation*: The authorization of another product or system to operate in a similar situation can be used to provide some assurance (e.g., reciprocity).

2. *Security Engineering*: Systems security engineering provides an elementary approach for building dependable systems in today's complex computing environment.

   2.1. *Planning and Assurance*: For new systems or for system upgrades, assurance requirements begin during the planning phase of the system life cycle.

   2.2. *Design and Implementation Assurance*: addresses a system's design as well as whether the features of a system, application, or component meet security requirements and specifications.

      2.2.1. *Use of Advanced or Trusted Development*: In the development of both commercial off-the-shelf (COTS) products and customized systems, the use of advanced or trusted system architectures, development methodologies, or software engineering techniques can provide assurance.

      2.2.2. *Use of Reliable Architecture*: Some system architectures are intrinsically more reliable, such as systems that use fault- tolerance, redundancy, shadowing, or redundant array of independent disks (RAID) features.

      2.2.3. *Use of Reliable Security*: One factor in reliable security is the concept of ease of safe use, which postulates that a system that is easier to secure is more likely to actually be secure. Security features may be more likely utilized when the initial system defaults to the "most secure" option.

      2.2.4. *Evaluations*: A product evaluation normally includes testing. Evaluations can be performed by many types of organizations, including: domestic and foreign government agencies; independent organizations such as trade and professional organizations; other vendors or commercial groups; or individual users or user consortia.

      2.2.5. *Assurance Documentation*: The ability to describe security requirements and how they were met can reflect the degree to which a system or product designer understands applicable security issues. Without a comprehensive understanding of the requirements, it is unlikely that the designer will be able to meet them.

2.2.6. ***Warranties, Integrity Statements, and Liabilities***: Warranties are an additional source of assurance. A manufacturer, producer, system developer, or integrator that is willing to correct errors within certain time frames or by the next release, gives the system manager a sense of commitment to the product and also speaks to the product's quality. An integrity statement is a formal declaration or certification of the product. It can be augmented by a promise to (a) fix the item (i.e., warranty) or (b) pay for losses (i.e., liability) if the product does not conform to the integrity statement.

2.2.7. ***Manufacturer's Published Assertions***: The published assertion or formal declarations of a manufacturer or developer provide a limited amount of assurance based on reputation. When there is a contract in place, reputation alone will be insufficient given the legal liabilities imposed on the manufacturer.

2.2.8. ***Distribution Assurance***: It is often important to know that software has arrived unmodified, especially if it is distributed electronically. In such cases, check bits or digital signatures can provide high assurance that code has not been modified. Antivirus software can be used to check software that comes from sources with unknown reliability (e.g., internet forum).

3. ***Operational Assurance***: addresses the quality of security features built into systems. Operational assurance addresses whether the system's technical features are being bypassed or have vulnerabilities and whether required procedures are being followed. It does not address changes in the system's security requirements, which could be caused by changes to the system and its operating or threat environment.

3.1. ***Security and Privacy Control Assessments***: Assessments can address the quality of the system as built, implemented, or operated. Assessments can be performed throughout the system development life cycle, after system installation, and throughout its operational phase. Assessment methods include interviews, examinations, and testing. Some common testing techniques feature functional testing (to see if a given function works according to its requirements) or penetration testing (to see if security can be bypassed).

3.2. **Audit Methods and Tools**: An audit conducted to support operational assurance examines whether the system is meeting stated or implied security requirements as well as system and organization policies. Some audits also examine whether security requirements are appropriate, though this is outside of the scope of operational assurance. Less formal audits are often called security reviews.

3.2.1. **Automated Tools**: Even for small multiuser systems, manually reviewing security features may require significant resources. Automated tools make it feasible to review even large systems for a variety of security flaws.

3.2.2. **Internal Controls Audit**: An auditor can review controls in place and determine whether they are effective. The auditor will often analyze both system and non-system based controls. Techniques used include inquiry, observation, and testing of both the data and the controls themselves. The audit can also detect illegal acts, errors, irregularities, or a lack of compliance with laws and regulations.

3.2.3. **Using the System Security Plan (SSP)**: The system security plan provides implementation details against which the system can be audited. This plan outlines the major security considerations for a system, including management, operational, and technical issues.

3.2.4. **Penetration Testing**: Penetration testing can use many methods to attempt a system break-in. In addition to using active automated tools as described above, penetration testing can be done "manually." The most useful type of penetration testing involves the use of methods that might be used against the system.

3.3. **Monitoring Methods and Tools**: Security monitoring is an ongoing activity that seeks out vulnerabilities and security problems. Many of the methods are similar to those used for audits but are done more regularly or, for some automated tools, in real time.

3.3.1. ***Review of System Logs***: A periodic review or use of automated tools to analyze system-generated logs can detect security problems, including attempts to exceed access authority or gain system access during unusual hours.

3.3.2. ***Automated Tools***: Several types of automated tools monitor a system for security problems. Some examples follow:

- ***Malicious code scanners*** are a popular means of checking for malicious code infections. These programs test for the presence of malicious code in executable program files;

- ***Checksum functions*** generate a mathematical value used to detect changes in the data based on the contents of a file. When the integrity of the file is being verified, the checksum is generated on the current file and compared with the previously generated value. If the two values are equal, the integrity of the file is verified. Running a checksum on programs can detect malicious code, accidental changes to files, and other changes to files. However, they may be subject to covert replacement by a system intruder. A digital signature, which guards against more than just accidental changes to files and are vastly superior to a checksum, can also be used to verify the integrity of a file;

- ***Password strength checkers*** test passwords against a dictionary (either a "regular" dictionary or a specialized one with easy-to-guess passwords, or both) and also check if passwords are common permutations of the user ID. Examples of special dictionary entries could be the names of regional sports teams and stars. Common permutations could be the user ID spelled backwards or the addition of numbers or special characters after common passwords;

- *Integrity verification programs* can be used by applications to look for evidence of data tampering, errors, and omissions. Techniques include consistency and reasonableness checks and validation during data entry and processing. These techniques can check data elements—as input or as processed—against expected values or ranges of values; analyze transactions for proper flow, sequencing, and authorization; or examine data elements for expected relationships. Integrity verification programs comprise a crucial set of processes meant to assure individuals that inappropriate actions, whether accidental or intentional, will be caught. Many integrity verification programs rely on logging individual user activities;

- *Host-based intrusion detection systems* analyze the system audit trail for activity that could represent unauthorized activity, particularly logons, connections, operating systems calls, and various command parameters; and

- *System performance monitoring* analyzes system performance logs in real time to look for availability problems, including active attacks, system and network slowdowns, and crashes.

3.3.3. *Configuration Management*: Configuration management provides assurance that the system in operation has been configured to organizational needs and standards, that any changes to be made are reviewed for security implications, and that such changes have been approved by management prior to implementation.

3.3.4. *Trade Literature/Publications/Electronic News*: In addition to monitoring the system, it is useful to monitor external sources for information. Such sources as trade literature, both printed and electronic, have information about security vulnerabilities, patches, and other areas that impact security.

4. *Interdependencies*: Assurance is an issue for every control and safeguard discussed in this publication. One important point to reemphasize here is that assurance is not only

for technical controls, but for operational controls as well. Although this chapter focused on systems assurance, it is also important to have assurance that management controls are working properly. Are user IDs and access privileges kept up to date? Has the contingency plan been tested? Can the audit trail be tampered with? Is the security program effective? Are policies understood and followed? The need for assurance is more widespread than individuals often realize.

Assurance is closely linked to planning for security in the system development life cycle. Systems can be designed to facilitate various kinds of testing against specified security requirements. By planning for such testing early in the process, costs can be reduced. Some kinds of assurance cannot be obtained without proper planning.

5. ***Cost Considerations***: There are many methods of obtaining assurance that security features work as anticipated. Since assurance methods tend to be qualitative rather than quantitative, they will need to be evaluated. Assurance can also be quite expensive, especially if extensive testing is done. It is useful to evaluate the amount of assurance received for the cost to make a best-value decision. In general, personnel costs drive up the cost of assurance. Automated tools are generally limited to addressing specific problems, but they tend to be less expensive.

System Support and Operations

System support and operations refers to all aspects involved in running a system. This includes both system administration and tasks external to the system that support its operation (e.g., maintaining documentation). It does not include system planning or design. The support and operation of any system—from a three-person local area network to a worldwide application serving thousands of users—is critical to maintaining the security of a system. Support and operations are routine activities that enable systems to function correctly. These include fixing software or hardware problems, installing and maintaining software, and helping users resolve problems.

The failure to consider security as part of the support and operations of systems, can be detrimental to the organization. Information Security system literature includes examples of how organizations undermined their often-expensive security measures with poor documentation, old user accounts, conflicting software, or poor control of maintenance accounts. An organization's policies and procedures often fail to address many of these important issues. Some major categories include:

- User support
- Software support
- Configuration management
- Backups
- Media controls
- Documentation; and
- Maintenance

Even though the goals of system support and operation and Information Security are closely related, there is a distinction between the two. The primary goal of system support and operations is the continued and correct operation of the system, whereas the Information Security goals of a system include confidentiality, availability, and integrity.

Security Control Families

To ensure the protection of confidentiality, integrity, and availability, minimum security requirements in multiple security-related areas should be required by an organization. The 20 control families introduced below represent a broad-based, balanced Information Security program that addresses the management, operational, and technical aspects of protecting information and systems.

Access Control (AC): The requirements for using—and prohibitions against the use of—various system resources vary considerably from one system to another. For example, some information must be accessible to all users, some may be needed by several groups or departments, and some may be accessed by only a few individuals. While users must have access to specific information needed to perform their jobs, denial of access to non-job-related information may be required. It may also be important to control the kind of access that is permitted (e.g., the ability for the average user to execute, but not change, system programs). These types of access restrictions enforce policy and help ensure that unauthorized actions are not taken.

Awareness and Training (AT): The user community is often recognized as being the weakest link in securing systems. This is due to users not being aware of how their actions may impact the security of a system. Making system users aware of their security responsibilities and teaching them correct practices helps change their behavior. It also supports individual accountability, which is one of the most important ways to improve Information Security.

Audit and Accountability (AU): An audit is an independent review and examination of records and activities to assess the adequacy of system controls and ensure compliance with established policies and operational procedures. An audit trail is a record of individuals who have accessed a system as well as what operations the user has performed during a given period. Audit trails maintain a record of system activity both by system and application processes and by user activity of systems and applications. In conjunction with appropriate tools and procedures, audit trails can assist in detecting security violations, performance issues, and flaws in applications.

Assessment, Authorization, and Monitoring (CA): A security control assessment is the testing and/or evaluation of the management, operational, and technical security controls on a system to determine the extent to which the controls are implemented correctly, operating as intended, and producing the desired outcome with respect to meeting the security requirements for the system. The assessment also helps determine if the implemented controls are the most effective and cost efficient solution for the function they are intended to serve. Assessment of the security controls is done on a continuous basis to support a near real-time analysis of the organization's current security posture.

Configuration Management (CM): Configuration management is a collection of activities focused on establishing and maintaining the integrity of information technology products and systems through the control of processes for initializing, changing, and monitoring the configurations of those products and systems throughout the System Development Life Cycle (SDLC). Configuration management consists of determining and documenting the appropriate specific settings for a system, conducting security impact analyses, and managing changes through a change control board. It allows the entire system to be reviewed to help ensure that a change made on one system does not have adverse effects on another system.

Contingency Planning (CP): An Information Security contingency is an event with the potential to disrupt system operations, thereby disrupting critical mission and business functions. Such an event could be a power outage, hardware failure, fire, or storm. Particularly destructive events are often referred to as "disasters." To avert potential contingencies and disasters or minimize the damage they cause, organizations can take early steps to control the outcome of the event. This activity is called contingency planning.

Identification and Authentication (IA): For most systems, identification and authentication is often the first line of defense. Identification is the means of verifying the identity of a user, process, or device, typically as a prerequisite for granting access to resources in a system. Identification and authentication are a technical measure that prevents unauthorized individuals or processes from entering a system.

Individual Participation (IP): Engagement with individuals whose information is being processed by a system is an important aspect of privacy protection and the development of trustworthy systems. System functions can have significant impacts on people's quality of life and their ability to be autonomous individuals. Effective engagement can help mitigate these risks and prevent a range of problems.

For example, individuals may feel surveilled by a system, which may create chilling effects on ordinary behavior or cause them to alter their interactions with the system in unexpected ways. They may feel information has been appropriated—or used for profit or organizational gain without their permission or sufficient economic benefit. Excluding access to information can affect data quality that could lead to adverse decision-making about users, including inappropriate restrictions on access to products or services or other types of discrimination.

Incident Response (IR): Systems are subject to a wide range of threat events, from corrupted data files to viruses to natural disasters. Vulnerability to some threat events can be mitigated by having relevant standard operating procedures that can be followed in the event of an incident. For example, frequently occurring events like mistakenly deleting a file can usually be repaired through restoration from the backup file. More severe threat events, such as outages caused by natural disasters, are normally addressed in an organization's contingency plan.

Maintenance (MA): To keep systems in good working order and to minimize risks from hardware and software failures, it is paramount that organizations establish procedures for the maintenance of organizational systems. There are many different ways an organization can address these maintenance requirements.

Media Protection (MP): Media protection is a control that addresses the defense of system media, which can be described as both digital and non-digital. Examples of digital media include: diskettes, magnetic tapes, external/removable hard disk drives, flash drives, compact disks, and digital video disks. Examples of non-digital media include paper or microfilm.

Privacy Authorization (PA): To better protect individuals' privacy and limit problems arising from system processing of their information, organizations should have a clear rationale for the collection, use, maintenance, and sharing of personally identifiable information (PII).

Overly broad collection and maintenance of information may create the potential for security vulnerabilities or allow for internal abuses or expanded uses that cross privacy boundaries. Individuals could be stigmatized by the release of their information or suffer from identity theft. Third parties with whom information is shared may disregard the purpose or context in which information is collected and use that information in a manner that contradicts individuals' privacy interests. As a result, individuals could lose trust in these systems, which could lead to abandonment or threaten the adoption of new technologies, even those designed to improve access to public services.

Physical and Environmental Protection (PE): The term physical and environmental security refers to measures taken to protect systems, buildings, and related supporting infrastructure against threats associated with their physical environment.

Planning (PL): Systems have increasingly taken on a strategic role in the organization. They assist organizations in conducting their daily activities and support decision making. With proper planning, systems can provide a security level commensurate with the risk associated with the operation of the system, improve productivity and performance, and enable new ways of managing and organizing. Planning for systems is crucial in the development and implementation of the organization's Information Security goals.

Program Management (PM): Systems and the information they process are critical to many organizations' ability to perform their missions and business functions. It makes sense that executives view system security as a management issue and seek to protect their organization's information technology resources as they would any other valuable asset. To do this effectively requires the development of a comprehensive management approach.

Personnel Security (PS): Users play a vital role in protecting a system as many important issues in Information Security involve users, designers, implementers, and managers. How these individuals interact with the system and the level of access they need to do their jobs can also impact the system's security posture. Almost no system can be secured without properly addressing these aspects of personnel security.

Risk Assessment (RA): Organizations are dependent upon information technology and associated systems to successfully carry out their missions. While the increasing number of information technology products used in various organizations and industries can be beneficial, in some instances they may also introduce serious threats that can adversely affect an organization's systems by exploiting both known and unknown vulnerabilities. The exploitation of vulnerabilities in organizational systems can compromise the confidentiality, integrity, or availability of the information being processed, stored, or transmitted by those systems.

System and Services Acquisition (SA): As with other aspects of information processing systems, security is most effective and efficient if planned and managed throughout a system's life cycle, from initial planning to design, implementation, operation, and disposal. Many security-relevant events and analyses occur during a system's life, which begins with the organization acquiring the necessary tools and services. The effective integration of security requirements into enterprise architecture also helps to ensure that important security considerations are addressed early in the SDLC and that those considerations are directly related to the organizational mission/business processes.

System and Communications Protection (SC): System and communications protection controls provide an array of safeguards for the system. Some of the controls in this family address the confidentiality and integrity of information at rest and in transit. The protection of confidentiality and integrity can be provided by these controls through physical or logical means. For example, an organization can provide physical protection by segregating certain functions to separate servers, each having its own set of IP addresses.

System and Information Integrity (SI): Integrity is defined as guarding against improper information modification or destruction and includes ensuring information non-repudiation and authenticity. It is the assertion that data can only be accessed or modified by the authorized personnel. System and information integrity provide assurance that the information being accessed has not been meddled with or damaged by an error in the system.

Making It Personal

Can you remember watching a crime drama on television or in the movies and hearing the phrase, "It's not personal. It's business"? For the cybercriminal, it's always business as usual. They don't give a hoot about who you are, what you do, or where you live. All they see is that you're a potential payday—a target to exploit. They want to strike quickly and covertly and walk away with your identity, money, or reputation. And truthfully, there isn't very much anyone can do about if they succeed because you may not ever be able to identify your attacker.

Up until now, this publication has explained much of the "whats" and "whys" of Cybersecurity. The "hows" and deeper details are for another discussion. Hopefully, a discussion that occurs after you hire us to implement your Information Security policies or audit the ones you already have in place; and we didn't even mention hardware or specific tools. That's also a different discussion.

Let's make it personal now. Regardless of how well protected or funded your organization is, you are not at work 24/7. And even if you are, a truly determined cybercriminal will persist at their break-in attempts until they find that one little flaw—that itsy-bitsy weakness in your protective shield that gives them just enough room to squeeze through and cause damage.

At home, you have personal computers, smart phones, social network accounts, and family members with the same, perhaps with even more devices than you. You don't have full-time cyber-guards protecting your personal assets. All of these devices and accounts are up for grabs if you don't protect yourself and your loved ones. You are probably at greater risk of having your personal systems compromised than your company's. You are vulnerable. You may have already been compromised without you even knowing it. Why don't you check to see if you've been pwned?

Have I Been Pwned?

Yes, "pwned" is spelled correctly. According to the Urban Dictionary, the word "pwned" is:

A corruption of the word "Owned." This originated in an online game called Warcraft, where a map designer misspelled "owned." When the computer beat a player, it was supposed to say, so-and-so "has been owned."

Instead, it said, so-and-so "has been pwned."

It means "to own" or to be dominated by an opponent or situation, especially by some god-like or computer-like force.

You can check whether you've been pwned and have an account that has been compromised in a data breach at https://haveibeenpwned.com/. At the time of this writing, Have I Been Pwned has details on 297 pwned websites, 5,369,804,192 pwned accounts, 75,653 pastes, and 82,644,754 paste accounts.

We haven't explained pastes, according to the Have I Been Pwned FAQs:

> Often when online services are compromised, the first signs of it appear on "paste" sites like Pastebin. Attackers frequently publish either samples or complete dumps of compromised data on these services.

A word to the wise—when you check if you have pwned accounts, try not to have anyone looking over your shoulder unless you are absolutely certain you won't be embarrassed by the search results. The website displays a list of every breach where your email has been compromised.

Is Your Password A Problem?

Every December since 2011, Internet security firm SplashData[15] has published its annual Worst Passwords List of the 25 most common passwords. Is your favorite password on the list?

Rank	Top 25 most common passwords by year according to SplashData						
	2011	2012	2013	2014	2015	2016	2017
1	password	password	123456	123456	123456	123456	123456
2	123456	123456	password	password	password	password	password
3	12345678	12345678	12345678	12345	12345678	12345	12345678
4	qwerty	abc123	qwerty	12345678	qwerty	12345678	qwerty
5	abc123	qwerty	abc123	qwerty	12345	football	12345
6	monkey	monkey	123456789	123456789	123456789	qwerty	123456789
7	1234567	letmein	111111	1234	football	1234567890	letmein
8	letmein	dragon	1234567	baseball	1234	1234567	1234567
9	trustno1	111111	iloveyou	dragon	1234567	princess	football
10	dragon	baseball	adobe123[16]	football	baseball	1234	iloveyou
11	baseball	iloveyou	123123	1234567	welcome	login	admin
12	111111	trustno1	admin	monkey	1234567890	welcome	welcome
13	iloveyou	1234567	1234567890	letmein	abc123	solo	monkey
14	master	sunshine	letmein	abc123	111111	abc123	login
15	sunshine	master	photoshop[17]	111111	1qaz2wsx	admin	abc123
16	ashley	123123	1234	mustang	dragon	121212	starwars
17	bailey	welcome	monkey	access	master	flower	123123
18	passw0rd	shadow	shadow	shadow	monkey	passw0rd	dragon
19	shadow	ashley	sunshine	master	letmein	dragon	passw0rd
20	123123	football	12345	michael	login	sunshine	master
21	654321	jesus	password1	superman	princess	master	hello
22	superman	michael	princess	696969	qwertyuiop	hottie	freedom
23	qazwsx	ninja	azerty	123123	solo	loveme	whatever
24	michael	mustang	trustno1	batman	passw0rd	zaq1zaq1	qazwsx
25	Football	password1	000000	trustno1	starwars	password1	trustno1

[15] http://www.splashdata.com/
[16] https://en.wikipedia.org/wiki/List_of_the_most_common_passwords#cite_note-fn1-11
[17] https://en.wikipedia.org/wiki/List_of_the_most_common_passwords#cite_note-fn1-11

If you see your password anywhere on this list, you've got one of the top worst passwords in the world. You can rest assured the Cybercriminals are laughing at you—all the way to the bank. You've just made their jobs super-easy.

We need to keep a lot of client passwords, so we can work on their projects. Even though we're super humans, it's impossible to remember the 1000s of passwords and accounts we have to track and manage on their behalf. It still amazes us that so many people use insecure passwords so it's easy for them to remember. If it's very easy for you to remember, then it's probably very easy for a Cybercriminal to breach.

Many of our clients only have one password that they use on all of their sites, and those passwords can usually be found in the pwned database. The problem with a single password approach is that you've just handed your attackers the keys to your kingdom. If they compromise just one account, they have access to all of your accounts.

Using a different unique, strong password for every account is one of the best ways to protect yourself even if an account is compromised. As an FYI, we always develop guidelines for strong password use in the Information Security policies we produce for organizations.

We use 1Password from AgileBits[18] in our business to manage all of our client accounts. 1Password integrates with your browser and Have I Been Pwned so you'll know immediately if you are using a compromised password. If you have an iPhone with IOS 12 or above, 1Password also integrates with mobile Safari. With 1Password, all it takes is 1-click to log into a site. AgileBits offers subscriptions for 5 family members at a very low cost.

---

[18] https://1password.com/

The Shadowy Case of Lucas Casimir

Despite taking every precaution, accounts can still be compromised in ways that we'll never be able to figure out as it happened to this author.

One morning, I awoke to an email and text message from Chase.com reporting that an old, infrequently accessed checking account was overdrawn by more than $1,700. Not knowing if this was a scam phishing attempt or a legitimate message, I logged into the account to discover two eCheck transfers to PayPal accounts totaling over $2,800.

Navigating through Chase's convoluted IVR system was daunting, but I finally reached a live person who helped freeze the account. Within an hour, a temporary credit was applied to the account effectively counteracting the cleared transaction.

Checking PayPal, there was no evidence that these transactions originated from my accounts.

Reaching a live person at PayPal is an unbelievably difficult task. An automated message said there would be at least a 3-hour wait before a real person is available. I followed the prompts to request a call back when someone breathing could talk to me.

PayPal's fraud investigator searched their database by bank account number. The investigation revealed that someone by the name of Lucas Casimir was using my bank account to make these transactions through his PayPal account. It's unknown if this is a real person, a false identity, or a stolen identity. Whatever the case, I'm completely unfamiliar with the name. It is not someone I know or ever heard of, but it is a name I shall never forget.

A deep scan of my laptop uncovered no malware, keyloggers, any other suspicious software. How Lucas got my bank account details remains a mystery although he may have purchased it on the dark web from either the JP Morgan Chase or OPM breaches.

If it happens to someone who is highly Cybersecurity aware, it can happen to anyone. Be careful and learn everything you can to protect yourself, your business, and your loved ones.

Use Antivirus Software and Keep It Updated

Antivirus software available today is quite sophisticated. All protect against viruses and most protect against ransomware, rootkits, trojan horses, and bots. Some even monitor WiFi traffic. It's a small investment to make to keep you from a world of hurt.

OS Firewalls

Some operating systems are delivered with built-in firewalls that monitor your ports for incoming and outgoing traffic. If your computer's operating system has a firewall, use it. You may be surprised by how many of your applications "call home".

Use a VPN

VPN means Virtual Private Network. If you use your computer on free WiFi networks available in public places like coffee shops, libraries, even co-located work spaces, you don't know if any security protocols are in place. A Cybercriminal with the right equipment can monitor and snoop on your activities. VPNs provide secure tunnels to the Internet encrypting your traffic, routing it though a server owned by the VPN company. With a VPN, nobody, not even the owner of the free Wi-Fi network, can snoop on your data.

Use 2-Factor Authentication

If you just use a password for authentication, anyone who learns that password owns your account. With two-factor authentication enabled, the password alone is useless. Two-factor authentication verifies your identity using at least two different forms of authentication: something you are, something you have, or something you know. It can be a bit of a pain to use, but it absolutely makes your accounts more secure.

Education and Awareness

Prior to closing its doors in 2011, the Syms Corporation had the slogan, "An Educated Consumer is our Best Customer". When it comes to protecting your family online, your education and awareness can make the difference between safeguard and misfortune. Don't click on links, even if they appear to be sent from people you know until you confirm their origin and intent. If you can't confirm, it may be a worm.

Take online courses or listen to a webinar or two. Educate yourself and bring security awareness to those you love.

Marketing Moment

I really hope you found this book to be educational, chock full of value, and helpful to you both personally and professionally. It's a lot to absorb but remember that Victor Font Consulting Group is here to help. Contact us at 1-844-VIC-FONT (842-3668) and we're happy to explore how we can be of service.

The Ultimate Guide to the SDLC

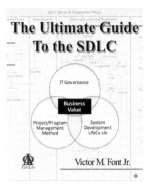

The term SDLC or System Development Life Cycle has been mentioned several times in this publication. The author has also written a comprehensive guide to the SDLC called, *"The Ultimate Guide to the SDLC"*, ISBN: 978-0985566647. You can purchase the book on Amazon[19] or on the author's website[20].

We cut our teeth in BIG IT and have worked for some of the largest and most successful organizations in the World. Even though our primary business has grown and evolved into the world-class digital strategy and software development house it is, we haven't abandoned our roots. We provide top-tier IT and CIO-for-rent consulting services on request.

Free Website/SEO Audit

Speaking of world-class digital strategy, your online business success is our passion! How is your online business website doing? Is it drawing enough traffic to be profitable? Or, does it scare people away?

Learn how you can outperform your competition, engage your visitors, and generate 30-50% or more online leads. Let us do a free, no obligation, website audit for you. Measure

---

[19] https://www.amazon.com/Ultimate-Guide-SDLC-Victor-Font/dp/0985566647/
[20] https://victorfont.com/shop/the-ultimate-guide-to-the-sdlc/

everything that matters to the online success of your business—*over 40 metrics in all!* Request your free website audit today[21].

No matter what your need, we've got you covered. Let us know how we can help by writing to info@victorfont.com

---

[21] https://victorfont.com/audit/